FAIRY TALES for a NEW FUTURE

Merryfield Oldsman
Illustrated by Susan Detwiler

Copyright © 2024 Merryfield Oldsman

All rights reserved.

No part of this book may be reproduced, stored in a retrieval system, or transmitted by any means, electronic, mechanical, photocopying, recording, or otherwise, without written permission from the author.

ISBN (Paperback): 979-8-9917098-3-5
ISBN (Hardcover): 979-8-9913611-6-3
ISBN (eBook): 979-8-9913611-5-6

The information provided in this book is for educational and informational purposes only and is not intended as medical advice. The use of mugwort tea and white pine needle tea can have varying effects on different individuals. It is important to consult with a qualified healthcare professional before consuming any herbal teas, especially if you are pregnant, nursing, have any existing health conditions, or are taking any medications.

The author and publisher of this book are not responsible for any adverse effects or consequences resulting from the use of any suggestions, preparations, or procedures described in this book. The reader assumes full responsibility for their actions and should always perform a patch test or consult with a healthcare provider to ensure safety and compatibility.

By reading this book, you acknowledge that you understand this disclaimer and that you use the information provided at your own risk.

Table of Contents

Fairy Tale	**Page**
Kyra, the Cave, and the Oak Box	1
Pearlina, the Heart, and the Willow	5
Tyler and the Mountain Ash	9
Ericka, the Lioness, and Herbs	13
Link, the Polar Bear, and the Olive Wand	17
Paxton and the Alder	23
Magdalena, the Yew, and the Ancestors	27
Erlinger, the Snake, and the Heartbeat of Earth	31
Eleanor, the White Pine, and the Peace Ceremony	35
Kai, the Birch, and the Earth Dance	39
The Sun, the Aspen, and the Circle of Children	43

Kyra, the Cave, and the Oak Box

Kyra was an explorer and adventurer. She lived near the edge of the glass and steel city, and it was easy for her to wander through the paved streets, skyscrapers on both sides, a strip of sky visible directly above, until she came to The Edge. At The Edge, the road stopped abruptly. The buildings cast long shadows onto the forest. There were many paths into and through the forest. Each path led to another path, to another path, to another path. Yet Kyra never got lost! Each time she arrived at a junction, she felt into her heart and her heart showed her the path to take. When returning to the city, she followed the same guidance, allowing her heart to return her to The Edge. The city. Home.

During her adventure time, after school and on weekends, Kyra discovered many hidden treasures in the forest. She found waterfalls. Some were little trickles of water over a few rocks in a little rill. Some were a hundred feet high with water from the Rain River rushing over the edge and down into a sparkling pool. She discovered hollowed-out oak trees with pockets large enough for her to stand in. She discovered bubbling streams with shiny mica and pyrite to pick up, hold in her palm, and feel the energies of. Splitting the mica, she delighted in a stone so sheer she could pull apart the layers and imagine all the years the mica spent in the water. The glitter of pyrite captivated her, and she held it, felt its cool energy, and admired its many facets.

Born when the Sun was in the Earth sign of Virgo, she had an affinity for rock and stone. One day, taking a new route through the forest, Kyra came to a cave, nestled into a rock wall. She cautiously peered into the darkness, wondering what might be within. She found a fist sized quartz stone on the path, picked it up, and tossed it into the cave as far as she could throw it. She heard a soft "thunk." The stone hitting the dirt ground! She knew the cave was deep since the quartz hit the ground and not the rear wall of the cave.

Kyra turned on the headlamp she kept with her for times her adventures brought her home after dark. She felt into her heart, asking if it was safe to enter the cave. She stood very still and said quietly, "I want to go into the cave and even more, I want what is best for me. Can I go into this cave?" She waited in the stillness and quiet. After a while, she noticed warmth and expansion in her chest around her heart. This was her knowing that yes, she could enter the cave. Slowly she moved forward, exploring the lumpy, yet smooth walls, the deep charcoal color, the emptiness within. No evidence of people or animals met her gaze. Feeling safe, she kept going. As she neared what she could see was the back of the cave, she saw a narrow passage off to her right. Using her lamp to peer within the passage, she saw it was only about twenty feet long and opened into another cave. A cavern! She slid sideways down the passage and stopped and stared when she entered the second cave. Moving her headlamp around, she saw enormous stalagmites rising from the ground in shades of soft white, yellow, and pale pink. Stalactites dripped from the ceiling in the same gorgeous hues. The space was so filled with these ancient structures that Kyra dared not walk further into the cavern for fear of damaging them.

There was a sudden rustle of movement and a lovely brown bat swooped about the cave, swishing close to Kyra, and then retreating. She held out her hands, appealing to the bat to come closer and as it approached her, blinked. The bat had transformed into a fairy, no bigger than Kyra's thumb! The fairy alit on Kyra's open palm and waved fairy fingers in the air. A dozen fairy friends flew from among the stalactites and perched on Kyra's shoulders and flitted around her head. Some gently brushed her hair from her eyes. Others danced in front of her, their sparkling bodies of light surrounding her.

Kyra felt her heart getting even bigger and warmer as the first fairy gently tapped Kyra's chest with her wand. "You are the first human to find our home in thousands of years. You are special and this cave is your special place. You can come here whenever you want, and you can bring a special friend with you. Even when you are not here, when you are in the forest or in the city, we are with you to protect you and keep you safe." With that, the fairy handed Kyra her wand, no bigger than a matchstick. Kyra accepted the gift with awe, sending love to all the fairies and receiving their love back. "How can I thank you?" Kyra whispered. "Your presence is enough," said the fairy as the other fairies shook and shimmered their agreement.

Kyra soaked in all the fairy light a little longer. Finally, when she had received and given all the love she could bring in and radiate out, she bid farewell to the fairies, knowing she would return soon.

She made her way back through the narrow passage and out of the first cave into the forest. There at the mouth of the cave she found a small oak box with a fairy figure delicately carved into the lid. In the center of the lid was embedded a cobalt blue sapphire. Kyra's birthstone! She knew in her heart that some magic had placed the oak box there for her as a container for her precious fairy wand. She picked up the box, opened the lid, and saw a velvet lining, the same cobalt blue as the sapphire. Carefully, she laid the wand inside the box, replaced the lid, and put the box in the zippered pocket of her pants. Feeling full of love and happiness, Kyra followed her heart back to the city.

PEARLINA, the HEART and the WILLOW

The Willow grew for many years before Pearlina found her. She had provided nesting branches for generations of cardinals and wrens. She had seen countless squirrels scamper through her supple limbs. She had witnessed numerous eclipses of the sun and moon. She had survived floods. She had known joys and sorrows as her neighbor trees lived and died. But she had never felt the joy that made her branches sway happily and her leaves shimmer with golden light the way they did when Pearlina arrived.

Pearlina wandered out of the wood and to the bank of the river. She was immediately drawn to the biggest, oldest Willow growing right at the river's edge, bending toward the flowing waters as if it, too, wanted to dive into the swirls and eddies. Pearlina approached the old tree slowly, looking at its deeply ridged bark. She felt Willow inviting her closer and she approached, tentatively at first, then joyously and lovingly as if she had become reacquainted with a long-lost friend. She hugged Willow and felt the tree energetically hugging her in return.

Climbing onto Willow's low hanging branches, Pearlina sat swinging her legs above the ground. She climbed higher and found squirrel and bird nests. Climbing even higher she discovered a knot on one of the branches. Peering inside the opening in the knot, she discovered a red crystal as big as her thumb. "Willow," she shouted, "Willow, how is there a crystal tucked into one of your branches?" She listened for an answer and heard only the rustle of leaves as the wind blew along the river. Pearlina did not know whether to return the crystal to its place in the tree or carry it with her, so she sat on the tree branch close to the trunk, placed the crystal over her heart, and became still. The stillness seemed to last forever until Pearlina heard a whisper in the leaves that softly told her to return the crystal to its home in the branch. The whisper continued, "Come back the night of the next full moon." Full moon thought Pearlina. I do not even know when that is. She climbed carefully down to the ground, thanked Willow with a hug and kiss and ran home to discover when the moon would next be full.

Two weeks later, Pearlina returned to the tree on the day of the full moon. She brought her sleeping bag and dinner so she could spend the night with Willow, Crystal, and the Moon. As the sun set, she ate her meal of fruit, vegetables, and nuts and drank her thermos of mugwort tea. She then toted her sleeping bag up into the halo of Willow's branches and waited for the rising moon. While waiting she practiced her favorite way of breathing that helped her to be patient. She breathed in while counting to five and breathed out counting to five. That pattern of soft breathing made her both relaxed and alert.

Seeing the moon rise above the horizon, Pearlina connected with the tree and heard, "It's time." She found her way to the knot on the branch and retrieved the crystal. It sparkled red, even in the moonlight.

Following her inner guide, she climbed back down to the ground with the crystal and carefully rinsed it in the river, right at the spot where the reflection of the moon on water kissed the shore. Putting the crystal on a special altar cloth she brought with her, she lay in the dewy grass with the river to her right and the crystal to her left. She took her moon bath, allowing the silvery energy of moonlight to bathe her with its magical, mysterious light.

As she moon-bathed, Pearlina felt a gently tingling begin in her toes and the soles of her feet. Slowly that tingling moved up through her legs, spreading within her torso, her arms, neck, and head. She felt light as if she could feel the spaces between the cells in her body. Light as if her density were shifting and becoming more crystalline. Again, she shifted to slower breathing and felt even more light, spacious, and free. All her cells were healthy and working perfectly. She took the crystal from the altar cloth and placed it on each chakra, breathing into her lower belly, moving the crystal to her upper belly, breathing, and repeating by placing the crystal on her heart, throat, and breathing into her body with each placement. As she placed the crystal on her forehead between her eyes, she saw a brilliant flash of rainbow light behind her eyelids. She realized she was seeing with her third eye! The rainbow flash connected her with her pineal gland deep within her brain and the pineal and pituitary glands connected and came online and she could see it all. She saw the river eons ago when it all began, all the trees that ever cast their shadows on this stretch of riverbank, all the animals that ever came to the stream to drink, and all the fish, crawdads, salamanders, and mud puppies that made the river their home. She heard the earthworms and insects in the earth and felt the power of the orbs in the heavens above. Most of all, she felt the roots of the Willow, watching her, guiding her, supporting her, grounding her as she experienced life on a new level.

Pearlina continued to watch the slow movement of the moon across the sky as the earth and heavens flowed within and around her. Willow gently brushed her leaves and branches across Pearlina's face, a sensation that felt like the breath of an angel to Pearlina.

As the moon reached its Zenith, Pearlina recovered her sleeping bag and gently replaced the crystal in its safe place. She curled up in her sleeping bag at the base of the willow and fell into a deep sleep. She dreamed that she was big. So big she contained all the Earth and the entire Universe within her body. She felt immense love for the whole of it, the oceans, mountains, people, and animals, the planets, sun, comets, and all life beyond Earth.

When Pearlina woke, the sun was up, Willow's branches were beside her head and Pearlina felt an inner stillness and peace that she had only imagined before her time with Crystal and the Willow.

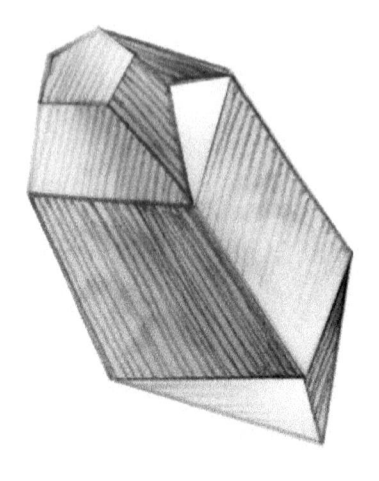

Tyler and the Mountain Ash

When Tyler walks, he likes to feel the sensation of his bare feet on the ground. He loves feeling the cool blades of grass, the sandy path, the gravelly road, all beneath his feet. Sometimes he wears sandals and feels the ground through the rubbery soles of his shoes. Sometimes he wears shoes or boots and even then, he feels the contact between the Earth and his feet. For Tyler, feeling the ground beneath him means he is fully alive. Thinking about what he will do next or what happened yesterday, or what he imagines might happen tomorrow all drift away on his ticker tape of consciousness, and he simply feels the earth beneath his feet.

He plays with walking. Sometimes he walks Tai Chi Style, placing first one heel on the ground, then rolling that foot down slowly until his toes touch the ground and then lifting almost onto his tiptoes while his other foot rises in the air. Then his other foot touches down heel first, rolls forward, up on tiptoes as the other foot rises to repeat the slow rhythmic walking.

Walking is one kind of meditation for Tyler. His favorite way to meditate is to walk through his yard until he comes to the Mountain Ash tree. He recognizes it by its slender leaves and sparkling berries. Sitting at the base of the Mountain Ash, Tyler lets his breath slow and deepen. He closes his eyes, safe and protected in the shadow of the tree. Turning inward in meditation with the Ash is magical to him. As Tyler breathes in and out, images form in his mind. Some stay with him for many breaths, others come in with the flow and out with the exhale. Tyler wants to share one of his meditations with you now. Here it is in his own words.

I sit with Ash, feeling the strength of its trunk running along my spine. For a moment, I am Ash Roots, deep underground, finding nourishment in the soil. Then, I am Ash Leaves, high atop the tree, soaking up the sunlight and dipping and rising on the windswept branches. Next Ash is in me, flowing Ash energy, watery, emotional, bringing images of dolphins and whales into my inner vision. One of the dolphins swims close and looks me directly in the eyes and I feel deep peace, the wisdom of the oceans, the knowing of infinity, in a momentary gaze. I feel one with it all, the entire universe within me, the oceans and mountains, plains and deserts, fields and streams, suns and moons, planets, and meteors, all within. Then all becomes light in and around me as the visions drift away and the one light remains. I sit in the one light, still, receptive, allowing, surrendering to its power. When I can hold no more of the deliciousness of pure light, my mind dives deep into my interior space, within my body. My mind takes me down to an awareness of my belly, dark and cavernous, lit from within, gently by the eternal flame that powers all the cells, all life. Within my belly, lit by the flame, I see the Inner Soul, primal, primitive, wild, and whole. "I've been waiting for you to return to me," Inner Soul says. I feel happy to see it, neither masculine nor feminine. Whole. The hieros gamos-the perfected blend of beingness. I sit with Inner Soul and see it is tending a fire towards the back of the cave within my belly. I move back toward the fire and Inner Soul and I sit together in stillness. We feel the appreciation of "as within, so without." As above, so below. "You know you are never alone," Inner Soul says silently with its eyes. "You have so many helpers and guides with you." We continue to sit together. Silently. For a long time. Then I feel a flutter in my heart and know it is time to go. Inner Soul knows it, too and walks with me towards the front of the cave within my belly. We say goodbye and I feel the transfer of that wild wholeness from Inner Soul to me. My imagination brings me back from Inner Focus to Outer Awareness and I feel Ash and my spine aligned. I mentally separate from Ash, pulling myself out of Ash and releasing Ash from within me. I stand and place my forehead against Ash's trunk in both blessing and gratitude for the meditation and visions we shared.

I walk home barefoot, feeling the grass and soil beneath my feet, grateful for my friend, Mountain Ash who brings me such fulfilling meditations.

As Tyler walks home, his pace is slow, peaceful, and purposeful. His mind is quiet, calm, and sharply aware of all that surrounds him. He thanks you for letting him share his favorite meditation with you!

Ericka, the Lioness, and Herbs

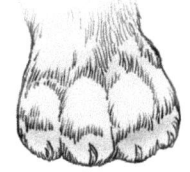

The mountain lion padded softly along the dusty trail, sniffing this way and that for her morning meal. She wanted something substantial. Bigger than the random mouse or hare. She wanted a fawn. Practiced in silent movement with a highly attuned sense of smell, she paced softly down the path.

Ericka was on the same path, out gathering morning herbs for the evening stew. She searched for wild thyme to add a touch of bitterness, burdock for earthy depth of flavor, and white pine for a boost to the immune system. Like the lioness, Ericka's keen sense of smell helped her find food. It also let her know who and what else was in the forest. She could tell when a fox was nearby when she smelled its musky scent. She picked up the scent of ants before coming upon their busy ant hill. She watched for scat that would warn her of bears and lions. She spent so much time in the forest that it was home to her. She deeply relaxed as she searched for the herbs she desired, while being open to whatever surprise the forest might share with her today. She knew she might encounter perfectly growing morels or hens of the woods. Or perhaps a patch of bluets, for a tiny bouquet. Dancing butterflies to witness. Some treasures returned home in her basket, others lived on in her memory, some treasured and others hidden until some similar event brought them alive again. On this morning, Ericka came to a patch of wild strawberries she didn't remember from past walks. Beside the strawberry patch was a flat, granite stone, perfect in size to serve as both table and chair to Ericka. She set down her basket upon the rock and began gathering strawberries into the pockets of her smock, delighting in the beautiful shades of red amongst the leafy green background.

As she popped a perfect berry in her mouth, her attention shifted. She noticed the slightest change in the breeze, in the air quality, in the scent in the air. She turned to face the path and saw the mountain lion crouched, looking at her, ready to race towards her and pounce upon her. Icy fear raced up Ericka's spine. Ericka breathed deeply into her fear, knowing that fear was her enemy in this situation. She knew she needed an ally. Quickly Ericka conjured up a being in her imagination. A woman. Strong and powerful. A woman with a big open heart, big enough to love and protect Ericka AND big enough to love and value the mountain lion. Ericka felt the woman behind her, towering over her, and then Ericka felt the woman's powerful love energy and physical strength merge with her own. She and the woman were now one. Ericka made eye contact with the mountain lion and saw the lioness shift from a pounce posture to one of alert curiosity. Ericka was connected with her own love. Her own power. Could the lioness sense that? She appeared to, as she sat on her haunches, relaxed and still alert. Ericka shifted her eyes lower, focusing on the lioness's chest, directing her own attention to the heart of the lioness. She then connected with her own heart, still beating rapidly and yet open to whatever might happen. Then Ericka saw an image of a silver, silken cord travelling from her own heart to that of the lioness. The lioness looked confused. Her hunger blended with peace in such a way that she no longer thought of food. This girl was working some sort of magic upon her so that she no longer wished to attack. She felt satisfied and content.

Ericka continued to connect to the big cat's heart and the cat continued to relax into alert aliveness. Ericka wondered how long her attention could hold. She was beginning to tire of the mental effort. She imagined once again the powerful woman ally who was at her back and immediately her focus returned, and Ericka continued to maintain her heart connection with the cat.

Leaves rustled behind the cat and both Ericka and the lioness looked into the woods. They saw a spotted fawn, separated from its mother. Ericka expected the cat to go after the fawn, but instead the lioness rose, looked at Ericka and then the fawn, and continued down the path at a slow, ambling pace.

Ericka sat on the rock watching her go and feeling grateful that she had trained herself to connect with her heart and develop love for all people, animals, plants, and all life. She knew her training served her well. She sent streams of gratitude to her guides and teachers who had trained her so lovingly. Ericka munched on a few more strawberries, gathered her basket and herbs, and slowly followed the lioness out of the woods and back home.

Link, The Polar Bear, and the Olive Wand

The cold winds blew and blew. It was spring, but the cold stayed longer than it was welcome. Link and Kyra had gone on a hike through the low mountains. Winds kicked up and icy rains began to fall before they were able to get home. As a cold rain saturated the earth and wind gusts blew limbs off dead trees all around them, Link and Kyra made their way to a rocky outcropping in the hopes of finding shelter amidst the boulders. Link carried with him a special amber crystal and an olive branch he used for dowsing. His crystal and wand had guided him through many situations and predicaments, and he trusted these sacred objects would help him now. The dowsing wand quivered each time Link pointed it north, so that is the direction they headed, moving swiftly towards the northern end of the Rocky Ridge. Kyra tuned into her own heart and knew north was the direction to go by the warmth she felt when she focused on the path north.

A slender maple, riddled with holes of a hungry woodpecker crashed to the ground fifty feet ahead of them. They pulled back and waited, listening for other signs of danger. Hearing none, they forged ahead, the rocks just a few hundred yards away. The wind gusted so loudly they only heard their own breath as they moved steadily forward.

When they reached the rocks, Link saw a crevice between several huge boulders. He shimmied through the narrow space and found a sheltered area where the boulders formed walls and ceiling. Shouting to Kyra to follow, she slipped between the rocks and into the shelter. The two children sank to the floor, glad to be out of the wind and rain. They sat for some time, catching their breath, and listening to the storm rage outside the grotto. They closed their eyes and lay down on the flat, rocky surface, grateful for a chance to rest and warm up.

After a few minutes they rose and began to explore the natural room they had been fortunate to discover. Large boulders and slabs of rock had shifted through the years, forming an oblong space about fifty feet by thirty feet. Looking up, they saw cracks of space between some of the boulders where drips of rainwater trickled through here and there.

In the center, they saw a circle of small stones with old, charred ashes, a remnant from some earlier person's fire pit. They moved about slowly, waiting for the winds and rains to abate. As they did so, they explored the contours of the rocks, first with their eyes, noticing the intricacies of the surface of the boulders and the variations in shades of gray from light limestone to deep charcoal. They placed their hands on the boulders, feeling the coolness. As they went round the cavern, exploring their surroundings, Link noticed an intentional shape, carved into the stone. Where the dips and bulges of the stone made natural contours suggestive of a great bear, someone had taken a sculpting hammer and traced the shape of a bear into the rock. Then the rock artist etched lines into the bear's fur, bringing out the whitish rock beneath the surface gray. "A polar bear," exclaimed Link, as he took a step back to examine the beauty of the art.

As Link stepped back, magic happened! The rock polar bear image became fuzzy and puffy like a cloud. The bear image transformed into a live polar bear right before their eyes! Kyra moved closer to Link, and he reached his arm towards her in a gesture of "It's okay." Link reached out his arm with palm up, making an open gesture to the polar bear who sat with its back against the wall, just below his etched form.

"Thank you," the Polar Bear said. "You can talk?," Link practically shouted. "Yes, and I can hear, too. No need to shout." The children and bear all giggled and Link said, "You seem friendly, but maybe you think we are food for you?" The bear laughed. "I have been trapped in this rock since the last Ice Age! I have not eaten in twelve thousand years and I am not about to start now that I have company. How did you find me?"

Link showed the bear his olive wood wand and amber crystal. He shared how the cold rain and blasting winds caused them to seek shelter and how the wand and crystal led them to this sheltered space wherein they discovered the bear art on the cave wall. Bear told Link how people came to this space many years ago and built the fire for warmth, but they never discovered the bear in the wall.

Link wondered why he was the one to discover the bear. Bear told him that in the old days, every person had a spirit animal to guide and strengthen them. When the ancestors felt scared or helpless, lonely, or sad, their spirit animal was always with them and helped them feel balanced and alive again. "Maybe I am your spirit animal," suggested Bear. "Sometimes I am the spirit animal for those born in the cold months of the year. When is your birthday?" "January," Link said excitedly. "You are my spirit animal! I like that." "Well, now that we have met, you know I can be gentle and you also know I am a polar bear and I can be fierce. Are these qualities you have or want to have?" "Yes," said Link, "I can be gentle and sometimes I do want to be fierce. Sometimes I know how to be these and sometimes I don't." "Well then," said Bear, "as we travel through life together, you can learn a lot from me and I can learn a lot from serving you. You are a strong way finder. You found your way, amidst a storm, to safety and shelter, and to your spirit animal. Always trust yourself. Always trust your highest self. Your highest self is one with God and will show you the way. And I am with you for strength and support."

With that, the Bear gently tapped Link's head and heart with his big bear paw. Link reached out to hold his hand, but Bear began to dissolve into a cloud again and floated back into the boulder. Link turned to Kyra who was still looking at the cave wall the bear had disappeared back into. She faced Link with wide eyes. "That really happened, didn't it?," said Link. Kyra nodded silently. It was then the children noticed the winds and rain had quieted and the forest sounds of dripping water and singing birds could be heard once again. As they turned to leave, Link felt a rock beneath his left foot. As he looked down, he saw not a rock, but a single bear claw. He gasped, picked it up and held it to his heart. "Thank you," he whispered. Then he tucked the claw into the leather pouch he wore about his waist in which he kept precious finds. He nodded to Kyra and they slipped back through the rocks into the wet shimmery forest and headed home.

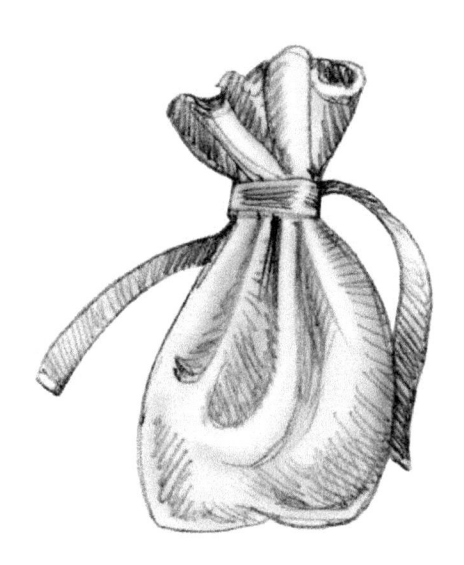

Paxton & the Alder

Paxton was born on a chilly, late December night when the new year of the Sun was just beginning in the land where the mountains and rivers meet. He grew strong and healthy, running barefoot through the woods and the fields where he learned to find autumn olives and Jerusalem Artichokes, wild blackberries, and burdock. His mama and dada showed him how to forage and fish and he was incredibly happy. In the summer of his 12th year, mother and father prepared a special dinner for him with all his favorite foods. Roasted trout, mashed potatoes, and asparagus from the gardens, and a special brew of Kombucha. After a dessert of pear crisp, his parents told him about the special plans to fulfill before his next birthday.

He would be given a special gift that lasts a lifetime and beyond. His very own birth tree. His parents led him to the truck in the driveway and there he saw a young tree, about six feet tall with its roots all wrapped up in a burlap ball. His parents were silent. They moved slowly and reverently as they approached the sapling. "What kind of tree is it?" Paxton wanted to know. He did not understand the importance of the tree, yet he sensed something sacred was happening by the demeanor of his parents. "It's an alder tree," his father replied quietly. "In the long-ago days, our ancestors felt their connection to the tree spirits even more than we do. They assigned a tree to time periods within the year. Everyone has their own tree based on their birthday. Your special birthday tree offers you strength, guidance, and support, just like our angel and animal guides. People born between December 27 and January 23 in the northern hemisphere have the Alder as their tree. This alder will teach you about trees and as you get to know your tree in a close and unique way, you will find a portal to knowing more about all the trees and all of life. The ancestors even created a tree alphabet called an ogham (pronounced ohm) that they used to understand big ideas about the universe. Now you are on the cusp of manhood and together we will plant your tree. Then you will take care of it and watch for the ways the tree takes care of you."

Paxton approached the tree and reached out, feeling its trunk, young, smooth, and greenish brown. He touched the young leaves, green tinged with red. "When can we plant it?" he asked. "Tomorrow," Dad said, unwrapping the burlap from the roots, lifting the tree, and carefully placing it in a washtub of cool water. "We will give the root ball time to soak in a lot of water, then we will take the tree down by the riverside."

The next day, Paxton could not wait to plant his tree. His mom and dad helped him dig a big hole, much bigger than the root ball. When it was just the right size. They put compost they had made from vegetable waste that sat in a barrel in the sun into the bottom of the hole. Next, they placed the tree in the hole. Paxton was strong enough to lift the tree and place it in the hole himself. Then mom and dad held it steady so he could fill in soil around the tree, covering the root ball and coming up the trunk.

Next, Paxton carried three bucketfuls of water to irrigate the roots. Mom and dad told him he would have to water it just that much every week unless there was a steady, soaking rain. Dad and Paxton got the wheelbarrow and wheeled four big bags of mulch from the truck to the tree and Paxton mulched the base of his tree to better hold the water in the soil.

Now it was time for Paxton to get to know his tree. He got paper and pencil and drew the leaves, so he knew their shape by heart. He compared the bark to other trees. He put his ear up to the tree to see if he could hear the water, sugar, and nutrients moving up and down within the trunk. Finally, he pressed his heart to the tree trunk asking Alder if their hearts could beat as one.

As the years passed, Paxton learned more about his Alder and himself. He asked it for strength when he got into arguments or became angry. He asked Alder for protection when he felt scared. And he asked the Alder for balance when he was not sure if he was being too meek or too forceful. He shared his joys with Alder when he caught a big fish, when he climbed a high mountain, and when he fell in love. Alder was always with him to help him know what he wanted and to help him in all ways.

Alder helped Paxton learn about all the trees around him. He learned to consult with birch when he was beginning a new project, willow when he wanted a burst of creativity, and apple when he wanted to feel his own abundance. Through his time with his Alder, Paxton learned more about himself. He learned to be quiet and listen. He learned patience. He learned determination to finish what was hard.

He learned to manage his emotions and turn sorrow into peace and joy. All through his life, wherever he went, Paxton's tree knowledge went with him. When he could, he returned to his first Alder, sat at its feet, and melted into the familiar comfort of its trunk.

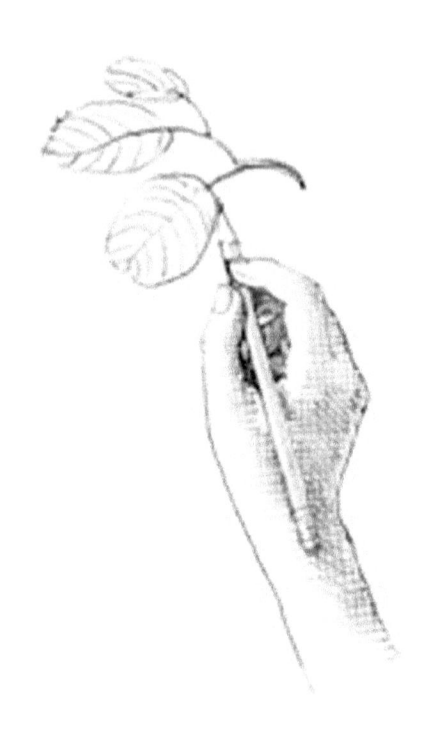

Magdalena, the Yew, and the Ancestors

Magdalena opened the iron garden gate and walked along the sandy trail until she reached her favorite place: The Garden Swing. Hung from strong oak timbers, the cool feel of the steel sent shivers through her body as she sat on the swing seat. Her leggings and flowery blouse were just enough clothing to keep her warm on this early spring day and still allow the cool breezes to stimulate and energize her. As she sat in the center of the swing, she raised her arms out sideways, breathed deeply, and felt the power of her own breath within and Gaia's breath all around her. She took several more breaths, in time with nature's breath, raised her arms overhead and looked up, eyes closed, basking in the sunlight.

Looking around the garden she felt the vibrations of her surroundings. The cedar elm, tall with a thick trunk and small leaves, waved to her in delight. She waved back and motioned from her heart to the tree, her language of connection and love. Next her gaze rested on the balsam cedar, its evergreen needles lightly bouncing in the breeze, welcoming, and inviting Magdalena into its world. A cluster of magnolias, leafless and not yet in bloom sent their sweet message to her and she nodded to their gray, twisty forms.

She looked upon the grasses, tall ornamental straw-colored grasses, green stubby grasses, all blanketing the areas between the walking paths. Unable to be simply an observer, Magdalena jumped from the swing, rolled in the grass, and lay still, contacting the heartbeat of Earth. Then she made the rounds, placing her hand, head, and heart on the trunk of each tree, feeling into its life energy within. Sometimes she placed her ear to the tree trunk and could hear the gentle whoosh of the tree's life fluids moving up and down through the trunk. Caressing the leaves and seed pods, she enjoyed her time with her flora friends.

The yew bush was her final stop. Gently touching its needle-like leaves, examining the twisty branches, Magdalena felt the call to be closer to the yew. Listening closely, she heard the yew inviting her on a journey into and through the tree. Finding a soft, grassy area nearby, she lay on the ground, feeling the earth below and the sky above. Slowing and deepening her breath, Magdalena imagined entering the branches of the yew and travelling down, down, down, through the woody trunk and deep into the roots. She was in total darkness as she dropped further beneath the earth's surface. The drop went on and on. She wondered when it would stop. There was no fear, only the gentle, ongoing pulses of the roots taking her deeper down. Finally with a gentle jolt, Magdalena's feet touched down and she crumpled in a heap on a hard packed dirt surface. The yew roots retracted, and she got to her feet, brushing dirt and pebbles off her clothes, and shaking out her hair. She turned and found herself standing at the entrance to a stone cave. A fire burned at the mouth of the cave and seated around the fire in a semi-circle facing her were The Ancestors. A half-dozen or more people, Neanderthal Ancestors, looked up at her in surprise. Their hair was uncut, men's faces unshaven, and they were wrapped in furry animal skins for warmth, though Magdalena did not feel cold. Low brows and heavy builds gave them the appearance of being not modern humans, but not hugely different.

They sat still, regarding her with ancient curiosity. She returned the interest, making eye contact with each of them in turn. As each connected with her, light rose in her eyes and in theirs. The Ancestors. Her Ancestors. Her Neanderthal Ancestors. They became emotional. Relief, gratitude, and grief all poured out of them as they wordlessly realized that she was their descendant. Magdalena was touched and felt their emotions and reflected gratitude back to them. She understood that they knew their time on Gaia was drawing to a close as the modern humans they met and bred with were dominating them. Until they saw Magdalena and recognized her, they thought they would soon be extinct, overcome by the dominant humans living among them. In meeting Magdalena, they knew they were not extinct, rather they were part of human evolution. Recognizing that Magdalena was their great-granddaughter many millennia in the future gave them deep peace.

Magdalena felt that same deep peace, for she too lived in a changing world where the survival of modern humans was uncertain. Just as her appearance satisfied the Neanderthals that they lived on through Magdalena, she realized that she, too, is an ancestor to those who will evolve in these current times. She joined their circle as they passed around a flaming stick, passing it over each of their hearts in an act of loving realization. No words or gestures could express the feelings of these precious, timeless moments.

Shortly, a rustle behind her reminded Magdalena that she had another time and place to return to. Bowing deeply to her beloved ancient ones, she turned and reached up, pulling herself into the root system of the yew. With a final smile and nod, the contemporary and ancients looked upon one another one final time, knowing they were forever connected in the Loving Heart of the Divine.

Erlinger and the Heartbeat of the Earth

Some days Erlinger is mad at the world. Some days he is a little bit mad, like when his sister plays with the toy he wants to play with. Some days he is a big angry mad like when he gets blamed for something he did not do. Some days he is a fuzzy mad with a slow current of anger washing this way and that through his belly and chest and fogging up his brain so he cannot think clearly. On those days, he does not even know why he is mad!

When Erlinger is a little mad, he wants to grab the toy back or hit his sister or punch the wall. When Erlinger is a big mad, he wants to scream and punch and kick and yell about how it is not right, and it is not fair! When Erlinger is fuzzy mad, he wants to hide away in his room, sleep, daydream, or stare out the window. He wants time and space to do nothing and wait for the fuzzy mad feelings to go away.

One day when Erlinger was in his room with his fuzzy anger, he looked out the window and saw a snake, coiled on a warm rock, sunning itself in the backyard. The snake looked so peaceful and calm, lying in the warmth of the sun. Erlinger wanted to have that peace and calm that he saw in the snake, so he went into the back yard, found a soft grassy area away from the snake but not too far, and laid down on his back in the soft grass.

Lying on his back with the sun kissing his face, Erlinger smelled the sweet, gentle scent of Mother Earth. He could identify the clean, deep, rich odor of black, earthy soil. He could pick out the lighter, gentler aroma of sweet grass. He noticed perfume scented air currents wafting above him and took in with delight honeysuckle and rose fragrances. Lifting his head, he noted the snake, a few yards away, still sunning itself on the rock, snake eyes closed, and he closed his eyes, too.

With his eyes closed, he heard more sounds around him. He heard the rustle of leaves blown by soft winds breezing through tree branches. He heard the crackle of birds hopping about in last autumn's fallen leaves. He heard the gurgle of the brook flowing just beyond the yard. He heard the chatter of a hawk swooping overhead, warning prey of its presence.

His attention then turned to the ground beneath him, supporting his weight. The texture of the earth was uneven, and a small rock gently poked his right shoulder. A roll of earth under his left knee lifted his leg slightly off the ground and a stick was quietly getting his attention underneath his head.

As he noticed all that he smelled, heard, and felt, Erlinger relaxed more deeply into the earth, beneath the sun, alongside the snake. He sensed the gentle beating of his heart, like a quiet rhythmic drum, thrumming within his chest. He rested deeper into the warm grasses. Briefly opening his eyes, he saw a wide expanse of blue sky above and a few puffy, grey-tinged white clouds traveling by. Closing his eyes and resting even deeper, he felt another heartbeat, outside of his body, coming up from the Earth beneath him. He put his attention on both heartbeats and as he listened, the heartbeats moved into sync with each other. His heart within his body. Mother Earth's heart deep within her body.

The two hearts connected as one and all Erlinger's anger flowed out, into the sky above and the earth below. As the anger surfaced, Erlinger saw light coming into his body-sunlight from above and earth-light from below. The two sources of light mixed in a dance of peace all through Erlinger's body. He lay still for a long time and fell asleep to the rhythmic dance of heart-filled love.

Waking up, he recalled he had been angry, but he could not even remember what he had been angry about! He stood up and noticed the snake had finished his nap, too, and had slithered off to someplace else. He ran into the house to tell his mom and dad and sister about the snake, his anger, and how good he felt exchanging heartbeats with Mother Earth.

His parents listened and decided that they, too, wanted to feel the earth's heartbeats. They decided to try lying on Mother Earth, too. And so, they did.

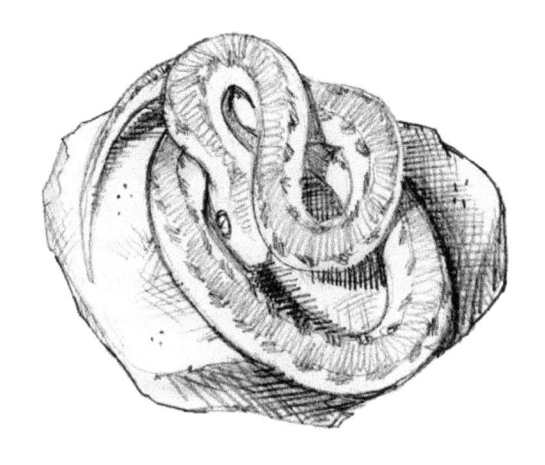

Eleanor, the White Pine, & the Peace Ceremony

When the world or her family was chaotic, Eleanor felt chaotic inside. Her heart beat faster, her legs bounced around shakily, and she could feel sweat beads rising on her body. Sometimes she felt tingles on the top of her head. When this happened, Eleanor would go to her room and lay in her bed and wait for the world inside to feel calm so she could face the chaotic world outside herself.

One sunny day in April, Eleanor's father and mother were talking loudly about something important to them, her older brother was slamming around noisily in the kitchen making his breakfast, and her younger brother was making fire engine noises while playing with his trucks. Eleanor did not even bother getting up. She lay in bed, listening to all the regular family noises and feeling jittery. She snuggled under her warm blanket, closed her eyes, and began to daydream. In her daydream, she lived in a quiet house with quiet people and felt calm inside.

A gentle tap on her shoulder startled her and she looked around. No one was in the room with her! Thinking it was a muscle spasm, Eleanor rolled over and got snuggly again. Then she felt the tap again and heard a voice say, "Sit up. I want to tell you something." The voice was not located outside her in the room. It also did not come through her mind as a thought. It was located someplace else that Eleanor did not understand and had not known! Yet it was a kind and gentle tone and Eleanor followed the direction and sat up in bed. She heard the voice again. "I am glad you are listening to me. I want to help you." Eleanor tuned in and listened. She was surprised at how well the voice knew her and understood her. She listened carefully as the voice, which she now considered a wise friend, gave her a skill she could use to create her own peace. The voice told her to go out to the grove of White Pine trees in her backyard. Find one pine, approach it, and sit at the base of the trunk. Once she was sitting at the pine, she was to imagine a special circle surrounding her and the pine. She could create the circle in any way she wanted. She could imagine herself surrounded by trees, or rock walls, or waterfalls. Whatever would allow her to feel safe and protected as she sat with her back to the tree inside her imaginary circle.

Eleanor got up and got dressed and quietly went outside. No one noticed her, and she was happy about that. She made her way through the gardens in the backyard and noticed one pine that beckoned her with waving branches. She chose that tree and sat against the trunk, noticing how rough the ground felt and how strong and sturdy the tree trunk felt as it supported her back and head. She breathed in the spicy pine scent. She imagined a thick grove of trees all around her, so thick she could no longer see her house or be seen from the house. She imagined a fire glowing in a fire pit in the center of her circle. She saw helpers and allies. People, animals, and trees all appeared in her imagination, supporting her, sending her peaceful energy, and keeping her safe.

She sat quietly, watching her breath move in and out. She felt her belly and chest rise and fall with each breath. She wondered what she was supposed to do next to create peace inside of herself. She noticed that doing even this much, she felt more peaceful. Leaning back a little more into the tree, she heard the tree whisper, "I am the tree of peace. Whenever you want peace, think of me. When you can, look at me. You can even carry a few of my needles in your locket necklace or carry one of my pinecones in your pocket."

As Eleanor sat listening to the tree, feeling its strength, smelling its warm scent, hearing the rustle of its branches, she felt peace growing inside her. "How do I keep this calm feeling when it's so noisy in my family?," Eleanor asked the tree. "You cannot always control what is going on outside of you," whispered the tree, "What you can do is take charge of how you think about it and what you do about it. As you practice peace within, you will be more heart-centered and more able to weather the storms and chaos happening around you."

"So, peace won't come to me in one magic moment?," asked Eleanor. "Not usually," replied the tree. "Sometime soon you will see and feel peace within and manifest it immediately. Until then, you get to return again and again to my peace offering to you. You can draw me, hug me, talk to me, even drink tea from my needles anytime you want to remember the peace you feel here with me now. The peace you feel now is yours. It is inside of you whether I am with you or not. As you grow your own peace and can call up a peaceful feeling in all situations, you become peace itself. Then you will know yourself as a free person, whatever happens around you."

Eleanor paused a moment as the white pine became silent. "I can't believe I'm talking to a tree!," she exclaimed. "Your higher self always knows what you need. When you listen to her voice, you reach another dimension where you can communicate with life around you in new ways." Hearing that, Eleanor felt grateful for her higher self, her physical self, and all of herself.

Gently moving away from the white pine, she saw a branch on the ground. She picked it up, said goodbye to the wise tree and carried the branch home to remind her that peace is always within her and it might take time, but she can always find it.

Kai, the Birch, and the Earth Dance

While Kai was sleeping, the sun, the moon, and the stars visited him. They came down into his sleeping room, entered his body, and washed through him as he soundly slept. The sun warmed his soul with its golden light. The moon cleared his mind with its silver glow. The stars activated the energies of all his cells with their multi-colored sparks of light and magical dust. As the sun, moon and stars visited, Kai dreamed he was a space traveler, soaring through the night heavens in a spaceship so light and clear that there was nothing between him and the entire universe. He swooped and floated, sped up and slowed down, all in the enchanting realm of space.

Eventually, Kai's dream ended. The sun, moon, and stars returned to their places in the sky, and Kai woke up, opening his eyes to golden rays of sunshine illuminating particles of dust wafting through his room. He felt lighter than before his nap, as if the density of his human body receded, allowing the light of his soul to move forward. This realization led Kai to want to dance. He got dressed and went outside barefoot. He stood on the dewy morning grass and wiggled his toes, getting to know the feel of the grass and earth beneath his feet. He looked up at the sun, rising over the mountain above the river. He felt the sun warming his face with the smile of life. Funny, he thought, I look at the sun every day and it feels different today! He jumped in the air, clapping his hands and began to dance his dance of the sun, lifting his hands and face, moving his arms as high, low, and side to side as far as he could. He felt soft air currents swirl around him as he danced. Looking up, he was surprised to see the waxing moon, smiling down on his dance. He rejoiced that the sun and moon were watching him together in the sky. He lay on his belly on the ground and rolled around. Then he rolled onto his back and lifted his arms and legs to the sky. Finally, he lay still, taking in the sweet warmth of life all around him.

As he lay still, he heard a whispery voice. Sitting up and looking around, he saw no one and lay back down. "You can't see me," he heard. "You can feel me in your heart. Are you feeling me in your heart?" Kai tuned in to the sturdy beating of his heart. As he did, an image of a tree came into his head. He saw a beautiful tree with three trunks, rising from a soft, marshy place in the ground. The trunk was silvery white with pinkish patches that were peeling or falling from the tree. "I know you," whispered Kai. You are the tree at the river bank where the water curves by the rocky cliff. "Yes," said the tree. "You're listening to me just right." "You are a birch, aren't you?," Kai asked. "I'm coming to visit you."

Kai made his way across the fields to the place where the birch tree lives. His heart was at peace as he approached the birch and reached out to touch its beautiful bark. Birch and Kai were silent for a long time, enjoying each other's presence and connecting heart to heart. In the silence, Kai noticed insects crawling along one of the birch branches. He saw the catkin flowers that sprouted in early spring. Looking up he saw a Crow, sitting high on a branch watching him. "Hello Crow," Kai called out softly. Crow cawed in response.

"Hey Birch and Crow, do you want to see my dance?," Kai asked excitedly. He heard a soft whisper of Yes from Birch and Crow flew to the ground and hopped around on one leg, a message Kai took to mean "Yes." So, Kai once again raised his face and arms to the sun and began dancing, his bare feet capturing a rhythm only he could hear. Or so he thought. As he danced, he saw birch begin to gently sway its branches in time with the music he heard in his head. He saw Crow begin to flap her wings and move her feet to that same rhythm. Kai began humming the tune aloud. Birch added a whispery swish swish in all the right places. Crow punctuated the tune with perfectly placed "caw caws". The child, Crow, and Birch shared the dance of life until the air filled with dancing birds and insects, the grasses, plants, and spring flowers swished and swayed. All the earth vibrated with the energy of life, filling all the spaces within and around everything. Fish jumped from the river; rocks pulsed with aliveness.

As everything swirled, danced, and vibrated, the Earth rose towards the heavens and the heavens descended to meet the earth. All the people all around the world noticed something just a little different as they went about their day. They all felt a little lighter and a little happier and a little more connected to the Earth below and the sun, moon, and stars above.

Kai's dance grew slower and finally paused. Even so, the beautiful life energy of the dance continued in him and all the people, plants, animals, minerals, and everything in, around, and above the earth. Kai raced home, excited to tell his mom and dad about the Earth Dance.

When they saw him coming across the field, they resonated with the joy he exuded and the whole family came outside barefoot to join him by dancing a few steps of his light-giving Earth Dance.

The Sun, the Aspen, and the Circle of Children

It was a warm summer day. The sun was high in the sky as the children ran from their homes and gardens and gathered in Aspen Grove. They held hands and formed a circle around the aspen sapling they had planted earlier in the year. Singing from their hearts they sang the Aspen Blessing to the trees.

> *Your grandmothers know our grandmothers.*
> *Your grandfathers know our grandfathers.*
> *We know you.*
> *We love you.*
> *You will meet our children.*
> *You will meet our grandchildren.*
> *We bless you.*
> *As you bless us.*

Round and round the children danced, singing the blessing again and again. When they grew warm with the heat of the sun, they lay down in the grass, fanning themselves out from the sapling, their heads faced inward towards the young aspen. Their bare feet, brown from the sun, calloused from the sands and stones they walked on, spread outwards towards the villages and woods.

The children felt at peace lying on their backs on the ground. They opened their eyes for a moment, allowing sun energy to kiss them through their eyes. Then they closed their eyes and felt the touch of the sun on their bellies, their hearts, their throats, and their foreheads. They allowed the warm caress of the sun to fill their legs, feet, and toes. They delighted in the warmth of the sun on their arms, hands, and fingers.

As all that warmth spread through their growing bodies, they felt the earth beneath them. They were attuned to the subtle movements of the earthworms, grubs, and other life forms beneath the grass in the soil. As they quieted and tuned in to Mother Earth, they felt her gentle heartbeat, ever powerful, ever loving, and ever giving. Their own Heartbeats began to synchronize with that of Mother Earth and with each other. As they lay together, on Mother Earth, fanned out around the young Aspen, they felt deep peace.

They had heard the stories of the before times. Before the earth was cherished. Before humans knew peace within themselves and around the world. They knew somewhere deep inside, without really knowing how they knew, that they were the ones bringing the healing. Their peace, their reverence, their love brought healing to the ancestors all the way back through time into the ancient past when the fear and wars began. They also knew those times were past and their practice of love and honor for Earth, trees, sun, moon, all of life and each other would bring peace and healing to the planet for future generations.

After about 20 minutes, moving as one, they rose, bowed to the ground, kissed the sapling, hugged each other, and broke into smaller groups to play.

One group made their way to the stream where they waded into the water, overturning rocks to find salamanders and crayfish whom they stroked and sang to.

One group shouted "tag" and a girl yelled, "I'm It," and chased a boy until she caught and tagged him, and he became "It" and continued the chase.

Another group played hide and seek, darting behind trees and scurrying into dens formed by boulders as their friends searched for them.

Another group gathered flowers for the community dining table, carefully picking flowers that would bloom a few more days and asking each if it was ready to serve the community with its bloom.

Slowly and easily, group by group, the children began to trickle back to the community of small homes arranged around the Big House. Each family had private living space and the Big House held the communal belongings and activities.

Parents and elders prepared the evening meal, enjoyed a cup of tea, tended the garden, cared for the chickens and goats, and worked peacefully, communicating with the plants and animals as they worked. They paused at their tasks, watching the children arriving back from the grove.

Some of the elders remembered the before times and how life was separate from nature and the Divine. They remembered the stress that led everyone, from infants to elders, to feel unsettled, separate, alone. They smiled as they considered the greater ease in a life connected to the Sun, Moon, Earth, plants, animals, and all of life.

The seasons were harsh. Physical labor was strenuous with hoeing, milking, harvesting, and preserving. Even so, living and working together, the people's relationships were strong and deep. Whatever challenges they faced, they knew they were there for each other, today, tomorrow, and always.

Milton Keynes UK
Ingram Content Group UK Ltd.
UKHW031442261124
451530UK00010B/100